A Level Mathematics
Lesson on C1 Differentiation

Here I offer a complete Lesson on Maths A level suitable to C1 Differentiation. This set of notes completely covers the chapter with 50 worked examples.

With over 20 years experience in teaching O level and A level maths (Pure mathematics and Mechanics) I offer this set of notes covering a wide range of problems with complete solutions. In this way I hope to help students achieve a high score in their A level maths examinations.

Each Lesson covers theory and formula necessary for the chapter and step by step explanation of all solutions. Problems are arranged in an ascending order of difficulty reaching A level standard.

The aim is to show students as many worked examples as possible which is not practical to see either in class or with a personal tutor.

All answers are provided(suggested)by the author.

Answers provided may not necessarily be the only possible solutions.

Attempt problems by yourself , then check the solution.

G. C. Ioannou
BSc(Hons),MSc,DIC

NB1(SOS)   **NOTE:**
y=k   NB=Notice Below
dy/dx=0   SOS=Memorise
k=constant

EXAMPLE 1   **EXAMPLE 2**   **EXAMPLE 3**
y=100   y=π   note: π=3.14...   y=-1/4
solution   solution   solution
dy/dx=0   dy/dx=0   dy/dx=0

NB2(SOS)   **NB3(SOS)**
y=x   y=kx
solution   solution   where k=constant
dy/dx=1   dy/dx=k

EXAMPLE 4   **EXAMPLE 5**   **EXAMPLE 6**
y=7x   y=x/6   y=-5/7x
solution   solution   solution
dy/dx=7   dy/dx=1/6   dy/dx=-5/7

NB4(SOS)   **NB5(SOS)**
y=x^n $1/x^n = x^{-n}$
solution
dy/dx= nx^{n-1}

EXAMPLE 7   **EXAMPLE 8**   **NB6(SOS)**
y=x^4 y=$1/x^5$ y=ax^n
solution   solution   solution
dy/dx=$4x^{4-1}$ y=x^{-5} dy/dx=anx^{n-1}
dy/dx=$4x^3$ dy/dx=$-5x^{-5-1}$
                    dy/dx=$-5x^{-6}$ where
                       or   a=constant
                    dy/dx=$-5/x^6$

EXAMPLE 9

Given that y=x

Show that dy/dx=1

<u>solution</u>

$y=x$

$y=x^1$

$dy/dx=1x^{1-1}$

$dy/dx=1.x^0$

$dy/dx=1.1$

$dy/dx=1$

EXAMPLE 10

$y=5/x^9$

<u>solution</u>

$y=5x^{-9}$

$dy/dx=5(-9)x^{-9-1}$

$dy/dx=-45x^{-10}$

or

$dy/dx=-45/x^{10}$

EXAMPLE 11

$y=4x^{12}$

<u>solution</u>

$dy/dx=4(12)x^{12-1}$

$dy/dx=4x^{11}$

NB7(SOS)

$\sqrt[n]{x^m}=x^{m/n}$

$\sqrt[n]{x}=x^{1/n}$

EXAMPLE 12

$y=\sqrt{x}$

<u>solution</u>

$y=x^{1/2}$

$dy/dx=1/2x^{1/2-1}$

$dy/dx=1/2x^{-1/2}$

$dy/dx=1/(2\sqrt{x})$

EXAMPLE 13

$y=5/\sqrt{x}$

<u>solution</u>

$y=5/x^{1/2}$

$y=5x^{-1/2}$

$dy/dx=5(-1/2)x^{-1/2-1}$

$dy/dx=-5/2x^{-3/2}$

EXAMPLE 14

$y=x^2-6x+7$

<u>solution</u>

$dy/dx=2x-6+0$

$dy/dx=2x-6$

note:

differentiate each term
separately based on
rules mentioned before.

EXAMPLE 15

$y=(x-4)^2$

<u>solution</u>

note: expand brackets first.

$y=(x-4)(x-4)$

$y=x^2-4x-4x+16$

$y=x^2-8x+16$

then differentiate

$dy/dx=2x-8+0$

$dy/dx=2x-8$

EXAMPLE 16
Find dy/dx for each of the following:

a) $y=x^3-x^8$ b) $y=3x^4-5$ c) $y=1/x^5-2/x$

<u>solutions</u>

a) $y=x^3-x^8$

$dy/dx=3x^{3-1}-8x^{8-1}$

$dy/dx=3x^2-8x^7$

b) $y=3x^4-5$

$dy/dx=3(4)x^{4-1}-0$

$dy/dx=12x^3$

c) $y=1/x^5-2/x$

$y=x^{-5}-2x^{-1}$

$dy/dx=-5x^{-5-1}-2(-1)x^{-1-1}$

$dy/dx=-5x^{-6}+2x^{-2}$

or

$dy/dx=-5/x^6+2/x^2$

EXAMPLE 17
Given that

$y=(3x-1)(2x+3)$

Find dy/dx

<u>solution</u>

First expand the brackets

$y=(3x-1)2x+3)$

$y=6x^2+9x-2x-3$

$y=6x^2+7x-3$

then differentiate

$dy/dx=12x+7$

EXAMPLE 18
Find the value of dy/dx at the point where $x=2$ on the curve whose equation is $y=(2x-3)(3x+1)$

<u>solution</u>

$y=(2x-3)(3x+1)$

first expand the brackets

$y=6x^2+2x-9x-3$

$y=6x^2-7x-3$

then differentiate

$dy/dx=12x-7$

when $x=2$

$dy/dx=12(2)-7$

$dy/dx=24-7$

$dy/dx=17$

Second Derivatives
Notations:

First derivative

$dy/dx=f'(x)$

Second derivative

$d/dx(dy/dx)=d^2y/dx^2$

or $f''(x)$

note: Differentiate the first derivative to get the second

Third derivative

$d/dx(d^2y/dx^2)=d^3y/dx^3$
or $f'''(x)$

note: Differentiate the second derivative to get the third etc

EXAMPLE 19

Given $y=x^3-7x^2+6x+3$
Find d^2y/dx^2
solution
$y=x^3-7x^2+6x+3$
$dy/dx=3x^2-14x+6$ (1)
$d^2y/dx^2=6x-14$ (2)
note: differentiate (1) to get (2)

EXAMPLE 20

Given that $y=x^{5/2}+3x^{7/2}$
find dy/dx and d^2y/dx^2
solution
$y=x^{5/2}+3x^{7/2}$
$dy/dx=5/2x^{5/2-1}+3(7/2)x^{7/2-1}$
$dy/dx=5/2x^{3/2}+21/2x^{5/2}$ (1)
$d^2y/dx^2=5/2(3/2)x^{3/2-1}+21/2(5/2)x^{5/2-1}$
$d^2y/dx^2=15/4x^{1/2}+105/4x^{3/2}$

EXAMPLE 21

Find dy/dx of
$y=\sqrt[3]{x^5}-5x^2+1/x^5-7$
solution
$y=x^{5/3}-5x^2+x^{-5}-7$
$dy/dx=5/3x^{5/3-1}-10x^{2-1}-5x^{-5-1}-0$
$dy/dx=5/3x^{2/3}-10x-5x^{-6}$
note:
$\sqrt[n]{x^m}=x^{m/n}$
$1/x^n=x^{-n}$

EXAMPLE 22

Find the gradient of the curve
$y=3x^2$ at the point(1,3)
solution
$y=3x^2$
$dy/dx=3(2)x^{2-1}$
$dy/dx=6x$
at (1,3)
when $x=1$
$dy/dx=6(1)$
$dy/dx=6$
or $m=6$
where m =gradient

EXAMPLE 23

Find the value of x if the gradient of the curve $y=x^2-5x$ is 3
solution
$y=x^2-5x$
$dy/dx=2x-5$
but $dy/dx=3$
$2x-5=3$
$2x=8$
$x=4$

Tangents and Normals to Curves

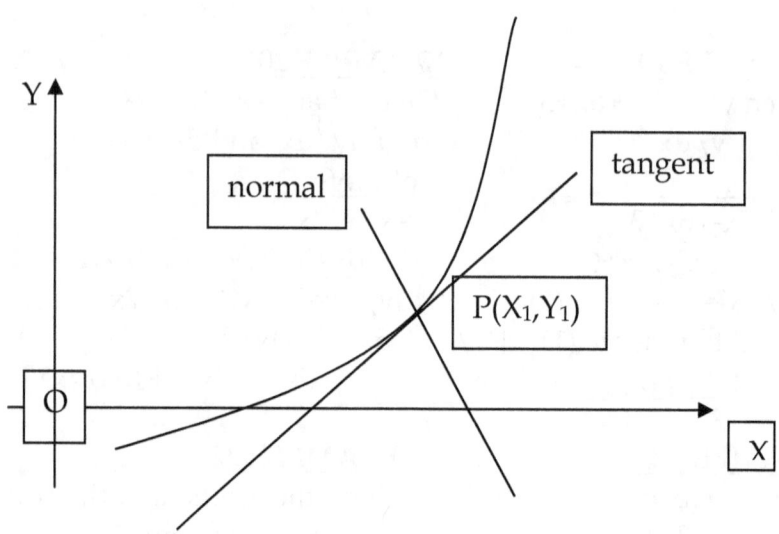

$m_t = dy/dx$ at the point $x = x_1$

note: substitute x in dy/dx to get the gradient of the tangent.

$\quad m_t.m_n = -1$ **(SOS)** for perpendicular lines.

$\quad$ or

$\quad m_n = -1/m_t$ **(SOS)**

where

$\quad m_t$ = gradient of the tangent

$\quad m_n$ = gradient of the normal

Equation of the tangent

$y - y_1 = m_t(x - x_1)$ **(SOS)**

Equation of the normal

$y - y_1 = m_n(x - x_1)$ **(SOS)**

EXAMPLE 24

Find the equation of the tangent and of the normal at the point P(2,16) on the curve with equation $y=2x^3$

solution

$y=2x^3$

$dy/dx=6x^2$

at the point P(2,16)

when x=2

$dy/dx=6(2)^2$

$dy/dx=24$

$m_t=24$

Equation of tangent

$y-y_1=m_t(x-x_1)$

$m_t=24$ P(2,16)

$y-16=24(x-2)$

$y-16=24x-48$

$y=24x-32$

Equation of normal

$m_n=-1/m_t$

$m_n=-1/24$

$y-y_1=m_n(x-x_1)$

$m_n=-1/24$ P(2,16)

$y-16=-1/24(x-2)$

note:

multiply by 24 to eliminate fractions

$24(y-16)=-1(x-2)$

$24y-384=-x+2$

$x+24y-386=0$

EXAMPLE 25

Find the equation of the tangent of the curve $y=x^2-7x+3$ at $x=1$

solution

$y=x^2-7x+3$

when $x=1$

$y=(1)^2-7(1)+3$

$y=1-7+3$

$y=-3$

$P(1,-3)$

$dy/dx=2x-7$

at the point $P(1,-3)$

when $x=1$

$dy/dx=2(1)-7$

$dy/dx=-5$

$m_t=-5$

Equation of tangent

$y-y_1=m_t(x-x_1)$

$m_t=-5$ $P(1,-3)$

$y-(-3)=-5(x-1)$

$y+3=-5x+5$

$y=-5x+2$

EXAMPLE 26

Find the gradient of the tangent and the gradient of the normal to the curve $y=x^2+5x-2$ at the point where $x=2$

solution

$y=x^2+5x-2$

$dy/dx=2x+5$

when $x=2$

$dy/dx=2(2)+5$

$dy/dx=4+5$

$dy/dx=9$

$m_t=9$ (gradient of tangent)

$m_n=-1/m_t$

$m_n=-1/9$ (gradient of normal)

EXAMPLE 27

Find the equation of the tangent and the normal to the curve $y=x^3+5x^2-2x-1$ at the point $(-3,23)$

solution

$y=x^3+5x^2-2x-1$

$dy/dx=3x^2+10x-2$

at the point $(-3,23)$

when $x=-3$

$dy/dx=3(-3)^2+10(-3)-2$

$dy/dx=27-30-2$

$dy/dx=-5$

$m_t=-5$

Equation of tangent

$y-y_1=m_t(x-x_1)$

$m_t=-5$ $(-3,23)$

$y-23=-5(x+3)$

$y-23=-5x-15$

$y=-5x+8$

Equation of normal

$m_n=-1/m_t$

$m_n=-1/-5$

$m_n=1/5$

$y-y_1=m_n(x-x_1)$

$m_n=1/5$ $(-3,23)$

$y-23=1/5(x+3)$

note: multiply by 5 to eliminate fractions

$5(y-23)=1(x+3)$

$5y-115=x+3$

$x-5y+118=0$

EXAMPLE 28

Given that $S=r^2(3r+1)(2-5r)$ find dS/dr

solution

$S=r^2(3r+1)(2-5r)$ note: first expand the brackets

$S=r^2(6r-15r^2+2-5r)$ and simplify

$S=r^2(-15r^2+r+2)$

$S=-15r^4+r^3+2r^2$ note: then differentiate

$dS/dr=-60r^3+3r^2+4r$

EXAMPLE 29

Given $y=3x^{1/2}-5x^{-1/2}$ find d^2y/dx^2

Solution

$y=3x^{1/2}-5x^{-1/2}$

$dy/dx=3(1/2)x^{1/2-1}-5(-1/2)x^{-1/2-1}$ note: apply the rules of

$dy/dx=3/2x^{-1/2}+5/2x^{-3/2}$ (1)   differentiation

                                                  correctly.

note: differentiate (1) to get the second derivative

$d^2y/dx^2=3/2(-1/2)x^{-1/2-1}+5/2(-3/2)x^{-3/2-1}$

$d^2y/dx^2=-3/4x^{-3/2}-15/4x^{-5/2}$

EXAMPLE 30

Differentiate the following:

     a) $y=4x^{-3}-5/(2x^4)+5\sqrt{x^3}$

     b) $f(x)=(3x^2-4)^2$

    solution

      a) $y=4x^{-3}-5/(2x^4)+5\sqrt{x^3}$

         $y=4x^{-3}-5/2x^{-4}+x^{3/5}$ note: change all to index form

       then differentiate

         $dy/dx=4(-3)x^{-3-1}-5/2(-4)x^{-4-1}+3/5x^{3/5-1}$

         $dy/dx=-12x^{-4}+10x^{-5}+3/5x^{-2/5}$

   note:

       $\sqrt[n]{x^m}=x^{m/n}$

        $1/x^n=x^{-n}$

b) $f(x)=(3x^2-4)^2$ note: first expand the brackets
   $f(x)=(3x^2-4)(3x^2-4)$
   $f(x)=9x^4-12x^2-12x^2+16$
   $f(x)=9x^4-24x^2+16$
note: then differentiate
   $f'(x)=9(4)x^{4-1}-24(2)x^{2-1}$
   $f'(x)=36x^3-48x$

EXAMPLE 31
Find the gradient of the curve $y=(3x-1)/x^5$ at $x=-2$
Solution
$y=(3x-1)/x^5$
$y=3x/x^5-1/x^5$
$y=3x^{-4}-x^{-5}$
$dy/dx=-12x^{-5}+5x^{-6}$
note: make powers of x positive then substitute values. Its easier.
$dy/dx=-12/x^5+5/x^6$
when $x=-2$
$dy/dx=-12/(-2)^5+5/(-2)^6$
$dy/dx=-12/-32+5/64$
$dy/dx=3/8+5/64$ note: LCM=64
$dy/dx=24/64+5/64$ LCM=Lowest Common
$dy/dx=29/64$ Multiple
or $m=29/64$ m=gradient

EXAMPLE 32

Prove that the tangent at (-2,-28) to the curve with equation $y=2x^3-3x^2$ is parallel to the line $y=36x-7$.
Find also the equation of the normal at (2,4) to the curve with equation $y=2x^3-3x^2$, giving your answer in the form $y=mx+c$

solution
$y=2x^3-3x^2$
$dy/dx=6x^2-6x$ note: for the line $y=36x-7$
at the point (-2,-28)   line in the form $y=mx+c$
when x=-2 $m_2=36$
$dy/dx=6(-2)^2-6(-2)$
$dy/dx=24+12$ note: condition for two lines to be
$dy/dx=36$ parallel $m_1=m_2$
$m_1=36$ (gradient of the tangent)
since $m_1=m_2$ lines are parallel
for the equation of normal:
$dy/dx=6x^2-6x$
at the point (2,4)
when x=2   note: when you substitute the x
$dy/dx=6(2)^2-6(2)$ value you in dy/dx you find
$dy/dx=24-12$ the gradient of the tangent.
$dy/dx=12$
$m_t=12$
$m_n=-1/m_t$
$m_n=-1/12$

Equation of normal
$y-y_1=m_n(x-x_1)$
$m_n=-1/12$ (2,4)
$y-4=-1/12(x-2)$
$y-4=-1/12x+1/6$
$y=-1/12x+4+1/6$
$y=-1/12x+25/6$

EXAMPLE 33

The tangent to the curve y=3x+2/x at the point where x=1 meets the axes at (a,0) and (0,b).Find the values of a and b.

solution

$y=3x+2/x$ (1)

$y=3x+2x^{-1}$

$dy/dx=3-2x^{-2}$

$dy/dx=3-2/x^2$

when x=1

$dy/dx=3-2/1^2$

$dy/dx=1$

$m_t=1$

substitute x=1 in (1)

$y=3(1)+2/1$

$y=3+2$

$y=5$

(1,5)

Equation of tangent

$y-y_1=m_t(x-x_1)$

$m_t=1$ (1,5)

$y-5=1(x-1)$

$y-5=x-1$

$y=x+4$

meets x-axis when y=0

$x+4=0$

$x=-4$ (-4,0)

meets y-axis when x=0

$y=4$

(0,4)

a=-4 b=4

EXAMPLE 34

The curve C has equation $y=2x^3-6x^2+7/x+10$, $x>0$

a) find dy/dx

b) show that the point P(1,13) lies on C

c) find an equation of the normal to C at the point P, giving answer in the form $ax+by+c=0$ where a, b and c are integers.

solution

a) $y=2x^3-6x^2+7x^{-1}+10$

$dy/dx=6x^2-12x-7x^{-2}$

$dy/dx=6x^2-12x-7/x^2$

b) P(1,13)

when x=1

substitute in $y=2x^3-6x^2+7/x+10$

$y=2(1)^3-6(1)^2+7/1+10$

$y=2-6+7+10$

$y=13$

curve passes through P

c) $dy/dx=6x^2-12x-7/x^2$

at the point P(1,13)

when x=1

$dy/dx=6(1)^2-12(1)-7/1^2$

$=6-12-7$

$=-13$

$m_t=-13$

$m_n=-1/m_t$

$m_n=-1/-13$

$m_n=1/13$

Equation of normal

$y-y_1=m_n(x-x_1)$

$m_n=1/13$ P(1,13)

$y-13=1/13(x-1)$

note: multiply by 13

$13(y-13)=1(x-1)$

$13y-169=x-1$

-x+13y-168=0

or

note: divide by -1

x-13y+168=0

EXAMPLE 35

Given that $f(x)=3x^3-9x^2-27x-20$

Find the points on the curve with equation $y=f(x)$, where the gradient is zero.

solution

$y=3x^3-9x^2-27x-20$ (1)

$dy/dx=9x^2-18x-27$

$dy/dx=0$ (gradient is zero)

$9x^2-18x-27=0$

note: divide by 9 to simplify the equation

$x^2-2x-3=0$

$(x-3)(x+1)=0$

$x=3$ or $x=-1$

when $x=3$ substitute in (1)

$y=3(3)^3-9(3)^2-27(3)-20$

$y=81-81-81-20$

$y=-101$

$(3,-101)$

when $x=-1$

substitute in (1)

$y=3(-1)^3-9(-1)^2-27(-1)-20$

$y=-3-9+27-20$

$y=-5$

$(-1,-5)$

EXAMPLE 36

Given that $y=6\sqrt{x}+2x^{1/3}+5$ find dy/dx and d^2y/dx^2

solution

$y=6\sqrt{x}+2x^{1/3}+5$

$y=6x^{1/2}+2x^{1/3}+5$

$dy/dx=6(1/2)x^{1/2-1}+2(1/3)x^{1/3-1}$

$dy/dx=3x^{-1/2}+2/3x^{-2/3}$ (1)

differentiating (1) again

$d^2y/dx^2=3(-1/2)x^{-1/2-1}+2/3(-2/3)x^{-2/3-1}$

$d^2y/dx^2=-3/2x^{-3/2}-4/9x^{-5/3}$

EXAMPLE 37

A Curve has equation $y=8+5x+4x^2-2x^3$. The gradient of the curve is -3 at the point P where $x=2$.Given that the tangent to the curve at the point Q is parallel to the tangent at P find the x-coordinate of Q.

solution

$dy/dx=-3$ (gradient at P)

$y=-2x^3+4x^2+5x+8$

$dy/dx=-6x^2+8x+5$

note: $m_p=-3$ and $m_p=m_Q$ since parallel

$-6x^2+8x+5=-3$

$-6x^2+8x+8=0$

note: divide equation by -2

$3x^2-4x-4=0$

$(3x+2)(x-2)=0$

$x=-2/3$ at Q

$x=2$ at P

EXAMPLE 38

Differentiate each of the following functions with respect to x.

a)$y=x^3+4x^2-16x+8$ b)$y=2x^{-5}+6/x-5/\sqrt{x}+x^4/x^3$

c)$y=(x^4-5x^6)/x$ d)$y=2\sqrt{x}(3x-1)$

solution

a) $y=x^3+4x^2-16x+8$

   $dy/dx=3x^2+8x-16$

b) $y=2x^{-5}+6/x-5/\sqrt{x}+x^4/x^3$ **NB(SOS):** $1/\sqrt{x}=1/x^{1/2}=x^{-1/2}$

   $y=2x^{-5}+6x^{-1}-5x^{-1/2}+x$ $1/x^n=x^{-n}$

   $dy/dx=2(-5)x^{-5-1}+6(-1)x^{-1-1}-5(-1/2)x^{-1/2-1}+1$

   $dy/dx=-10x^{-6}-6x^{-2}+5/2x^{-3/2}+1$

c) $y=(x^4-5x^6)/x$

   $y=x^3-5x^5$

   $dy/dx=3x^2-25x^4$

d) $y=2\sqrt{x}(3x-1)$

   $y=2x^{1/2}(3x-1)$

   $y=6x^{3/2}-2x^{1/2}$

   $dy/dx=6(3/2)x^{3/2-1}-2(1/2)x^{1/2-1}$

   $dy/dx=9x^{1/2}-x^{-1/2}$

EXAMPLE 39

A Curve has equation $y=x^3-6x^2+10x$.

     a)Show that the curve only crosses the x-axis at **only one** point.

The point P on the curve has coordinates (3,3).

     b)Find the gradient of the tangent to the curve at P and hence find the equation for the normal to the curve at P, giving your answer in the form y=mx+c.

The normal to the curve at P meets the coordinate axes at M and N.

     c)Find the area of the triangle OMN where O is the origin.

solution

a) $y=x^3-6x^2+10x$ (1)

   crosses x-axis when y=0

   $x^3-6x^2+10x=0$

   $x(x^2-6x+10)=0$

   x=0 or $x^2-6x+10=0$

                note: **Discriminant : D=b^2-4ac (SOS)**

$x^2-6x+10=0$

$a=1$ $b=-6$ $c=10$ note: If $D<0$ no real roots. Does not crosses x-axis.

$D=b^2-4ac$

$D=(-6)^2-4(1)(10)$

$D=36-40$

$D=-4<0$ No real roots.

Therefore crosses x-axis at only one point, that is when x=0.

When x=0 substitute in (1) : y=0

At the point (0,0)

b)$dy/dx=3x^2-12x+10$

   at the point P(3,3)

   when x=3

   $dy/dx=3(3)^2-12(3)+10$

   $\quad\quad =27-36+10$

   $\quad\quad =1$

   $m_t=1$

   Equation of normal

   $m_n=-1/m_t$

   $m_n=-1/1$

   $m_n=-1$

   $y-y_1=m_n(x-x_1)$

   $m_n=-1$ P(3,3)

   $y-3=-1(x-3)$

   $y-3=-x+3$

   $y=-x+6$

c)equation of normal: y=-x+6

   i) crosses x-axis when y=0

      $0=-x+6$

      $x=6$

      M(6,0)

   ii) crosses y-axis when x=0

      $y=6$

      N(0,6)

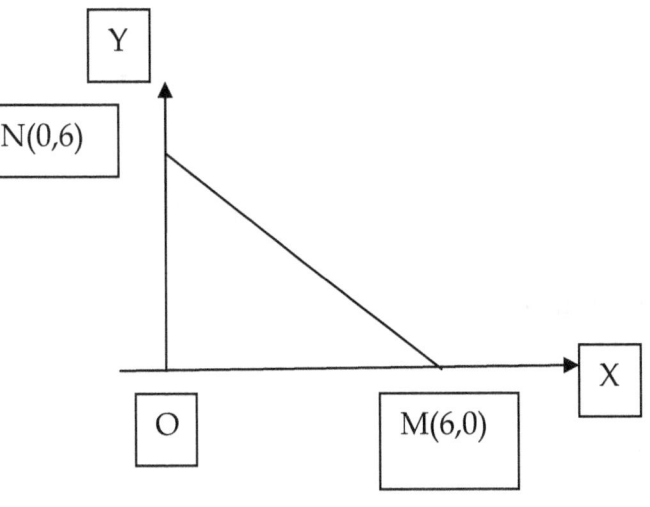

Area of triangle OMN = (b.h)/2

$$=(OM.ON)/2$$
$$=(6 \times 6)/2$$
$$=18 \text{ square units}$$

EXAMPLE 40

The curve C has equation $y=6x^2+(7-x)/x$, $x \neq 0$.
The point P on C has x-coordinate 1.
a) Show that the value of dy/dx at P is 5.
b) Find the equation of the tangent to C at P.
The tangent meets the x-axis at the point (m,0).
c) Find the value of m.

solution

a) $y=6x^2+(7-x)/x$ (1)

$y=6x^2+7/x-1$

$y=6x^2+7x^{-1}-1$

$dy/dx=12x-7x^{-2}$

when x=1

$dy/dx=12(1)-7/(1)^2$

$=12-7$

$dy/dx=5$

b) when x=1,
   substitute in (1)
   $y=6(1)^2+(7-1)/1$
   $y=6+6$
   $y=12$
   $P(1,12)$
   Equation of tangent
   $y-y_1=m_t(x-x_1)$
   $m_t=5$ $P(1,12)$
   $y-12=5(x-1)$
   $y-12=5x-5$
   $y=5x+7$

c) equation of tangent: $y=5x+7$
   meets x-axis when $y=0$
   $5x+7=0$
   $5x=-7$
   $x=-7/5$
   $(-7/5,0)$
   $m=-7/5$

EXAMPLE 41

A curve C has equation $y=2x^3-7x+4/x, x\neq0$.
The points A and B lie on C and have coordinates (2,4) and
(-2,-4).

   a) Show that the gradient of C at A is equal to the gradient
      of Cat B.
   b) Find an equation for the normal of C at A.

The normal to C at A meets the y-axis at the point M. The
normal to C at B meets the y-axis at the point N.

   c) Find the length of MN.

solution

   a) $y=2x^3-7x+4/x$
       $y=2x^3-7x+4x^{-1}$
       $dy/dx=6x^2-7-4x^{-2}$

$dy/dx=6x^2-4/x^2-7$

at the point A(2,4)

when x=2

$dy/dx=6(2)^2-4/2^2-7$

$dy/dx=24-1-7$

$dy/dx=16$

$m_A=16$

At the point B(-2,-4)

 when x=-2

$dy/dx=6(-2)^2-4/(-2)^2-7$

$\qquad =24-1-7$

$dy/dx=16$

 $m_B=16$

 $m_A=m_B$

b)Equation of normal at A

 $m_t=16$

 $m_n=-1/m_t$

 $m_n=-1/16$

 $y-y_1=m_n(x-x_1)$

 $m_n=-1/16$ A(2,4)

 $y-4=-1/16(x-2)$

 note: multiply by 16 to eliminate fractions

 $16(y-4)=-1(x-2)$

 $16y-64=-x+2$

 $x+16y-66=0$ (1)

c)Normal at A meets y-axis when x=0

 substitute x=0 in (1)

 $16y-66=0$

 $16y=66$

 $y=66/16$

 $y=33/8$

 M(0,33/8)

 Equation of normal at B

 $y-y_1=m_n(x-x_1)$

 $m_n=-1/16$ B(-2,-4)

$y+4=-1/16(x+2)$
note: multiply by 16 to eliminate fractions
$16(y+4)=-1(x+2)$
$16y+64=-x-2$
$x+16y+66=0$ (2)
 Normal at B meets y-axis when x=0.
Substitute x=0 in (2)
$16y+66=0$
$16y=-66$
$y=-66/16$
$y=-33/8$
N(0,-33/8)

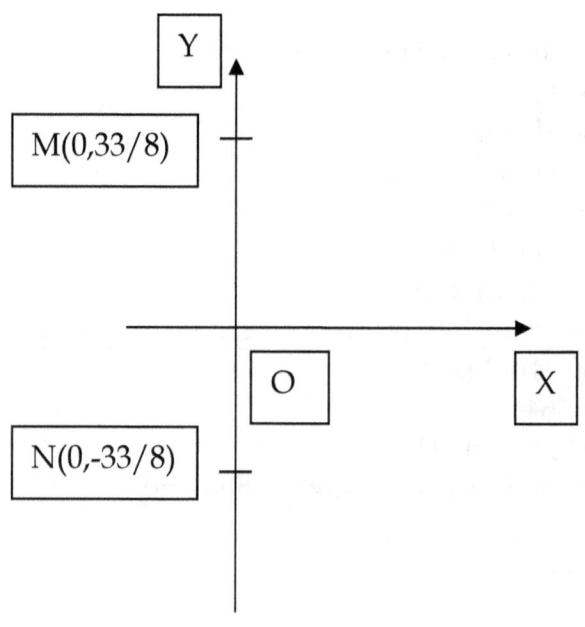

$MN=2\times33/8$
$MN=66/8$
$MN=33/4$
$MN=8\ 1/4$ Units

EXAMPLE 42

Given that $y = x^6 + (2+x^5)/\sqrt[3]{x}$

   a) show that y simplifies to $x^6 + 2x^{-1/3} + x^{14/3}$

   b) find dy/dx

solution

a) $y = x^6 + (2+x^5)/\sqrt[3]{x}$

  $y = x^6 + (2+x^5)/x^{1/3}$

  $y = x^6 + 2/x^{1/3} + x^5/x^{1/3}$

  $y = x^6 + 2x^{-1/3} + x^{5-1/3}$ note: $a^m/a^n = a^m : a^n = a^{m-n}$

  $y = x^6 + 2x^{-1/3} + x^{14/3}$

b) $dy/dx = 6x^{6-1} + 2(-1/3)x^{-1/3-1} + 14/3x^{14/3-1}$

  $dy/dx = 6x^5 - 2/3x^{-4/3} + 14/3x^{11/3}$

EXAMPLE 43

The curve C has equation $y = x^2 + 3x - 4$ and crosses the x-axis at the points A(a,0) and B(b,0) where a<b.

   a) show that the normal to the curve at A has equation
     x-5y+4=0.

The tangent to the curve at B meets the normal to the curve at A at the point K.

   b) Find the exact coordinates of K.

solution

   a) $y = x^2 + 3x - 4$

      crosses x-axis when y=0.

      $x^2 + 3x - 4 = 0$

      $(x+4)(x-1) = 0$

      x=-4 or x=1

      but a<b

      a=-4 and b=1

      A(-4,0) and B(1,0)

      $y = x^2 + 3x - 4$

      $dy/dx = 2x + 3$

      at A(-4,0)

      $dy/dx = 2(-4) + 3$

      $dy/dx = -8 + 3 = -5$

$m_t = -5$

$m_n = -1/m_t$

$m_n = -1/-5$

$m_n = 1/5$

Equation of normal

$y - y_1 = m_n(x - x_1)$

$A(-4,0) \quad m_n = 1/5$

$y - 0 = 1/5(x+4)$

$y = 1/5(x+4)$

note: multiply all by 5 to eliminate fractions

$5y = 1(x+4)$

$5y = x+4$

$x - 5y + 4 = 0 \quad (1)$

b) $dy/dx = 2x+3$

at $B(1,0)$

when $x = 1$

$dy/dx = 2(1)+3$

$dy/dx = 2+3$

$dy/dx = 5$

$m_t = 5$

Equation of tangent

$y - y_1 = m_t(x - x_1)$

$m_t = 5 \quad B(1,0)$

$y - 0 = 5(x-1)$

$y = 5x - 5 \quad (2)$

for point K solve simultaneously (1) and (2)

$x - 5y + 4 = 0 \quad (1)$

$y = 5x - 5 \qquad (2)$

substitute (2) in (1)

$x - 5(5x-5) + 4 = 0$

$x - 25x + 25 + 4 = 0$

$-24x + 29 = 0$

$-24x = -29$

$x = -29/-24$

$x = 29/24$

substitute $x = 29/24$ in (2)

y=5(29/24)-5
y=145/24-5   note: LCM=24
y=145/24-120/24
y=25/24
K(29/24,25/24)

EXAMPLE 44
Given that $y=(3x^6+2\sqrt{x})/x^3$ find:
  a)dy/dx b)d²y/dx²

solution
a) $y=(3x^6+2\sqrt{x})/x^3$
   $y=3x^6/x^3+2\sqrt{x}/x^3$
   $y=3x^3+2x^{1/2}/x^3$
   $y=3x^3+2x^{1/2-3}$
   $y=3x^3+2x^{-5/2}$
   $dy/dx=3(3)x^{3-1}+2(-5/2)x^{-5/2-1}$
   $dy/dx=9x^2-5x^{-7/2}$ (1)
 b) $d^2y/dx^2=9(2)x^{2-1}-5(-7/2)x^{-7/2-1}$
    $d^2y/dx^2=18x+35/2x^{-9/2}$ (2)
    note: differentiate (1) to get (2)

EXAMPLE 45
Given that $y=3x^2+6\sqrt{x}$
   a) find dy/dx
   b) show that $2xd^2y/dx^2+dy/dx-18x=0$
solution
 a) $y=3x^2+6\sqrt{x}$
    $y=3x^2+6x^{1/2}$
    $dy/dx=6x+6(1/2)x^{1/2-1}$
    $dy/dx=6x+3x^{-1/2}$ (1)
 b) $d^2y/dx^2=6+3(-1/2)x^{-1/2-1}$
    $d^2y/dx^2=6-3/2x^{-3/2}$ (2)
    note: differentiate (1) to get (2)
    substitute (1) and (2) into

L.H.S$=2xd^2y/dx^2+dy/dx-18x$

$\quad=2x(6-3/2x^{-3/2})+6x+3x^{-1/2}-18x$

$\quad=12x-3x^{-1/2}+6x+3x^{-1/2}-18x$

$\quad=0$

$\quad=$R.H.S

where

$\quad$ L.H.S=Left Hand Side

$\quad$ R.H.S=Right Hand Side

EXAMPLE 46

a) Given that $f(x)=(x+2)(x-3)^2$ find $f'(x)$

b) Given that $y=(x^2-12x-12)/(6x^{1/2})$
   show that dy/dx can be expressed in the form
   $(x+k)^2/(mx^{2/3})$ where k and m are integers to be found.

solution

a) $f(x)=(x+2)(x-3)^2$

$\quad$ note: first expand the brackets

$\qquad (x-3)^2=(x-3)(x-3)$

$\qquad\qquad =x^2-3x-3x+9$

$\qquad\qquad =x^2-6x+9$

$\quad f(x)=(x+2)(x^2-6x+9)$

$\quad f(x)=x^3-6x^2+9x+2x^2-12x+18$

$\quad f(x)=x^3-4x^2-3x+18$

$\quad$ then differentiate

$\quad f'(x)=3x^2-8x-3$

b) $y=(x^2-12x-12)/(6x^{1/2})$

$\quad y=x^2/(6x^{1/2})-12x/(6x^{1/2})-12/(6x^{1/2})$

$\quad y=x^{3/2}/6-2x^{1/2}-2x^{-1/2}$

$\quad dy/dx=3/2(1/6)x^{3/2-1}-2(1/2)x^{1/2-1}-2(-1/2)x^{-1/2-1}$

$\quad dy/dx=1/4x^{1/2}-x^{-1/2}+x^{-3/2}$

$\quad$ note: then factorize

$\quad dy/dx=1/4x^{-3/2}(x^2-4x+4)$

$\qquad =1/4x^{-3/2}(x-2)^2$

$\qquad =1/(4x^{3/2})(x-2)^2$

$\qquad =(x-2)^2/(4x^{3/2})$

$\quad$ where: k=2 and m=4

EXAMPLE 47

Water is poured into a tank such that the depth h cm, of the water in the tank after t seconds is given by $h=2kt^{2/3}$, where k is a constant. Given that when t=1, the depth of the water is increasing at the rate of 4cm per second
 a) find the value of k.
 b) find the rate at which is increasing when t=27

solution
 a) $h=2kt^{2/3}$
   when t=1 sec dh/dt=4cm/sec
   $dh/dt=2k(2/3)t^{2/3-1}$
   $dh/dt=(4k/3)t^{-1/3}$
   when t=1 sec dh/dt=4cm/sec
   $4=(4k/3)(1)^{-2/3}$
   $4=4k/3$
   $12=4k$ note: the word rate means differentiation
   $k=3$ i.e dh/dt
 b) $dh/dt=(4k)t^{-1/3}$
   when k=3
   $dh/dt=4(3)/3t^{-1/3}$
   $dh/dt=4t^{-1/3}$
   when t=27 sec
   $dh/dt=4(27)^{-1/3}$
   $dh/dt=4/27^{1/3}$
      $=4/\sqrt[3]{27}$
   dh/dt=4/3 cm/sec

EXAMPLE 48

A curve has an equation y=2x+9/x+5.
   a) Find the gradient of the curve at the point P(1,16).
The tangent to the curve at the point Q is parallel to the tangent to the curve at P.
   b) Find the coordinates of Q.

solution

a) $y=2x+9/x+5$ (1)

$\quad y=2x+9x^{-1}+5$

$\quad dy/dx=2-9x^{-2}$

$\quad dy/dx=2-9/x^2$

$\quad$ at the point P(1,16)

$\quad$ when x=1

$\quad dy/dx=2-9/1^2$

$\qquad\quad =2-9$

$\qquad\quad =-7$

$\quad m_p=-7$

b) $dy/dx=-7$ note: $m_p=m_Q$ since parallel

$\quad 2-9/x^2=-7$

$\quad -9/x^2=-9$

$\quad -9=-9x^2$

$\quad x^2=1$

$\quad x=+/-\sqrt{1}$

$\quad$ x=1 or x=-1

$\quad$ at P: x=1

$\quad$ at Q: x=-1

$\quad$ when x=-1

$\quad$ substitute in (1)

$\quad y=2(-1)+9/-1+5$

$\quad y=-2-9+5$

$\quad y=-6$

$\quad$ Q(-1,-6)

EXAMPLE 49

The curve C has an equation $y=2x^2+(3-2x)/x$, $x\neq0$.
The point P on C has an x-coordinate -1.

    a) show that the value of dy/dx at P is -7.

    b) find the equation of the tangent to C at P.

The tangent meets the x-axis at the point (m,0).

    c) find the value of m.

<u>solutions</u>

a) $y = 2x^2 + (3-2x)/x$ (1)

   $y = 2x^2 + 3/x - 2x/x$

   $y = 2x^2 + 3x^{-1} - 2$

   $dy/dx = 4x - 3x^{-2}$

   $dy/dx = 4x - 3/x^2$

   when $x = -1$

   $dy/dx = 4(-1) - 3/(-1)^2$

       $= -4 - 3$

   $dy/dx = -7$

b) when $x = -1$

   substitute in (1)

   $y = 2(-1)^2 + [3 - 2(-1)]/-1$

   $y = 2 + 5/-1$

   $y = 2 - 5$

   $y = -3$

   $P(-1, -3)$

   Equation of tangent

   $y - y_1 = m_t(x - x_1)$

   $m_t = -7$ $P(-1, -3)$

   $y + 3 = -7(x + 1)$

   $y + 3 = -7x - 7$

   $y = -7x - 7$

   $y = -7x - 10$

c) $y = -7x - 10$

   cuts x-axis when $y = 0$

   $-7x - 10 = 0$

   $-7x = 10$

   $x = -10/7$

   $(-10/7, 0)$

   $m = -10/7$

EXAMPLE 50

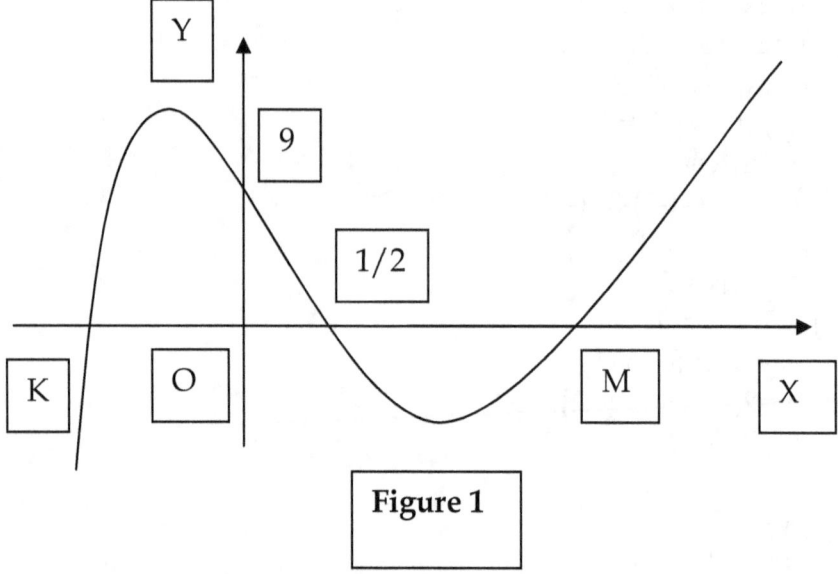

Figure 1

Figure 1 above shows part of the curve C with equation
y=(2x-1)(x²-9).
The curve meets the x-axis at the points, K,(1/2,) and M.
   a) write the x-coordinate of K and the x-coordinate of M.
   b) find dy/dx
   c) show that y=-14x+6 is the equation of the tangent at
      the point (1,-8).
The tangent to C at the point R is parallel to the tangent at the
point (1,-8).
   d)   Find the exact coordinates of R.

solution
a) y=(2x-1)(x²-9)   (1)
      crosses x-axis when y=0.
      (2x-1)(x²-9)=0
      (2x-1)(x-3)(x+3)=0
      x=1/2 or x=3 or x=-3
       at K: x=-3   at M: x=3

b) $y=(2x-1)(x^2-9)$
    $y=2x^3-18x-x^2+9$
    $y=2x^3-x^2-18x+9$
    $dy/dx=6x^2-2x-18$

c) $dy/dx=6x^2-2x-18$
    at the point (1,-8)
    when x=1
    $dy/dx=6(1)^2-2(1)-18$
           $=6-2-18$
    $dy/dx=-14$
    $m_t=-14$
    Equation of tangent
    $y-y_1=m_t(x-x_1)$
    $m_t=-14$ (1,-8)
    $y+8=-14x+14$
    $y=-14x+6$

d) $dy/dx=6x^2-2x-18$
    $dy/dx=-14$ at (1,-8)
    $m_1=m_2$ since parallel
    $6x^2-2x-18=-14$
    $6x^2-2x-4=0$
    note: since an equation divide by 2 to simplify
    $3x^2-x-2=0$
    $(3x+2)(x-1)=0$
    $x=-2/3$ at R and x=1
    when x=-2/3 substitute in (1)
    $y=[2(-2/3)-1][(-2/3)^2-9]$
    $y=(-4/3-1)(4/9-9)$
    $y=(-4/3-3/3)(4/9-81/9)$
    $y=(-7/3)(-77/9)$
    $y=539/27$

    R(-2/3,539/27)

    END OF LESSON!